YOUR KNOWLEDGE HAS VALUE

- We will publish your bachelor's and
  master's thesis, essays and papers

- Your own eBook and book -
  sold worldwide in all relevant shops

- Earn money with each sale

Upload your text at www.GRIN.com
and publish for free

C. P. Kumar

Groundwater Data Requirement and Analysis

GRIN Verlag

Bibliografische Information der Deutschen Nationalbibliothek:

Die Deutsche Bibliothek verzeichnet diese Publikation in der Deutschen National-
bibliografie; detaillierte bibliografische Daten sind im Internet über http://dnb.d-
nb.de/ abrufbar.

Imprint:

Copyright © 2014 GRIN Verlag GmbH
Druck und Bindung: Books on Demand GmbH, Norderstedt Germany
ISBN: 978-3-656-75553-1

This book at GRIN:

http://www.grin.com/en/e-book/281602/groundwater-data-requirement-and-ana-
lysis

GROUNDWATER DATA REQUIREMENT AND ANALYSIS

C. P. Kumar
National Institute of Hydrology
Roorkee – 247667 (Uttarakhand)

1.0   INTRODUCTION

Groundwater is used for a variety of purposes, including irrigation, drinking and manufacturing. Groundwater is also the source of a large percentage of surface water. To verify that groundwater is suited for its purpose, its quality can be evaluated (i.e., monitored) by collecting samples and analyzing them. In simplest terms, the purpose of groundwater monitoring is to define the physical, chemical, and biological characteristics of groundwater.

Accurate and reliable groundwater resource information (including quality) is critical to planners and decision-makers. Huge investment in the areas of groundwater exploration, development and management at state and national levels aims to meet the groundwater requirement for drinking and irrigation and generates enormous amount of data. We need to focus on improved data management, precise analysis and effective dissemination of data.

Numerical models are capable of solving large and complex groundwater problems varying widely in size, nature and real life situations. With the advent of high speed computers, spatial heterogeneities, anisotropy and uncertainties can be tackled easily. However, the success of any modelling study, to a large measure depends upon the availability and accuracy of measured/recorded data required for that study. Therefore, identifying the data needs of a particular modelling study and collection/monitoring of required data form an integral part of any groundwater modelling exercise.

2.0   DATA REQUIREMENT FOR GROUNDWATER STUDIES

The first phase of any groundwater study consists of collecting all existing geological and hydrological data on the groundwater basin in question. This will include information on surface and subsurface geology, water tables, precipitation, evapotranspiration, pumped abstractions, stream flows, soils, land use, vegetation, irrigation, aquifer characteristics and boundaries, and groundwater quality. If such data do not exist or are very scanty, a program of field work must first be undertaken, for no model whatsoever makes any hydrological sense if it is not based on a rational hydrogeological conception of the basin. All the old and newly-found information is then used to develop a conceptual model of the basin, with its various inflow and outflow components.

A conceptual model is based on a number of assumptions that must be verified in a later phase of the study. In an early phase, however, it should provide an answer to the important question: does the groundwater basin consist of one single aquifer (or any lateral combination of aquifers) bounded below by an impermeable base? If the answer is yes, one can then proceed to the next phase: developing the numerical model. This model is first used to synthesize the various data and then to test the assumptions made in the conceptual model. Developing and testing the numerical model requires a set of quantitative hydrogeological data that fall into two categories:

- Data that define the physical framework of the groundwater basin
- Data that describe its hydrological stress

These two sets of data are then used to assess a groundwater balance of the basin. The separate items of each set are listed below.

Physical framework

1. Topography
2. Geology
3. Types of aquifers
4. Aquifer thickness and lateral extent
5. Aquifer boundaries
6. Lithological variations within the aquifer
7. Aquifer characteristics

Hydrological stress

1. Water table elevation
2. Type and extent of recharge areas
3. Rate of recharge
4. Type and extent of discharge areas
5. Rate of discharge

It is common practice to present the results of hydrogeological investigations in the form of maps, geological sections and tables - a procedure that is also followed when developing the numerical model. The only difference is that for the model, a specific set of maps must be prepared. These are:

- Contour maps of the aquifer's upper and lower boundaries
- Maps of the aquifer characteristics
- Maps of the aquifer's s net recharge
- Water table contour maps

Some of these maps cannot be prepared without first making a number of auxiliary maps. A map of the net recharge, for instance, can only be made after topographical, geological, soil, land use, cropping pattern, rainfall, and evaporation maps have been made.

The data needed in general for a groundwater flow modelling study can be grouped into two categories: (a) Physical framework and (b) Hydrogeologic framework (Moore, 1979). The data required under physical framework are:

1. Geologic map and cross section or fence diagram showing the areal and vertical extent and boundaries of the system.
2. Topographic map at a suitable scale showing all surface water bodies and divides. Details of surface drainage system, springs, wetlands and swamps should also be available on map.
3. Land use maps showing agricultural areas, recreational areas etc.
4. Contour maps showing the elevation of the base of the aquifers and confining beds.

5. Isopach maps showing the thickness of aquifers and confining beds.
6. Maps showing the extent and thickness of stream and lake sediments.

These data are used for defining the geometry of the groundwater domain under investigation, including the thickness and areal extent of each hydrostratigraphic unit.

Under the hydrogeologic framework, the data requirements are:

1. Water table and potentiometric maps for all aquifers.
2. Hydrographs of groundwater head and surface water levels and discharge rates.
3. Maps and cross sections showing the hydraulic conductivity and/or transmissivity distribution.
4. Maps and cross sections showing the storage properties of the aquifers and confining beds.
5. Hydraulic conductivity values and their distribution for stream and lake sediments.
6. Spatial and temporal distribution of rates of evaporation, groundwater recharge, surface water - groundwater interaction, groundwater pumping, and natural groundwater discharge.

3.0   GROUNDWATER DATA ACQUISITION

Obtaining all the information necessary for modelling is not an easy task. In fact, a modeller may have to devote considerable effort and time in data acquisition, especially when the database for the study area is non-existent. Some data may be obtained from existing reports of various agencies/departments, but in most cases, additional field work is required. Moreover, the data is not readily available in the format required by the model, and requires additional work to process it. The observed raw data obtained from the field may also contain inconsistencies and errors. Before proceeding with data processing, it is essential to carry out data validation in order to correct errors in recorded data and assess the reliability of a record. In addition, as the modelling exercise progresses, certain gaps in the database get identified. In such cases, the field monitoring program may undergo some revision including installation of new piezometers/monitoring wells.

Amongst the hydrologic stresses including groundwater pumping, evapotranspiration and recharge, groundwater pumpage is the easiest to estimate. Field information for estimating evapotranspiration is likely to be sparse and can be estimated from information about the land use and potential evapotranspiration values. Recharge is one of the most difficult parameters to estimate. Recharge refers to the volume of infiltrated water that crosses the water table and becomes part of the groundwater flow system. This infiltrated water may be a certain percentage of rainfall, irrigation return flow, seepage from surface water bodies etc., depending upon topography, soil characteristics, depth to groundwater level and other factors.

Values of transmissivity and storage coefficient are usually obtained from data generated during pumping tests and subsequent data processing. For modelling at a local scale, values of hydraulic conductivity may be determined by pumping tests if volume-averaged values are required or by slug-tests if point values are desired. For unconsolidated sand-size sediment, hydraulic conductivity may be obtained from laboratory permeability tests using permeameters. However, due to rearrangement of grains during repacking the sample into permeameter, the obtained hydraulic conductivity values are typically several orders of magnitude smaller than values measured *in situ*. Furthermore, in a laboratory column sample, large-scale features such as fractures and gravel lenses that may impart transmission characteristics to the hydrogeologic unit as a whole are not captured. Due to this reason, laboratory analyses of core samples tend to give lower values of hydraulic conductivity than are measured in the field. In the absence of site-specific field or laboratory measurements, initial estimates for aquifer properties may be taken from standard tables.

When simulating anisotropic media, information is required on principal components of hydraulic conductivity tensor K_x, K_y and K_z. Vertical anisotropy is defined by the anisotropy ratio between K_x and K_z. For most groundwater problems, it is impossible to model geologic units at the isotropic scale. When the thickness of the model layer (B_{ij}) is much larger than the thickness of isotropic layer (b_{ijk}) (assuming this thickness can be identified from bedding information), the hydrologically equivalent horizontal and vertical hydraulic conductivities for the model layer may be calculated as:

$$(K_x)_{ij} = \sum_{k=1}^{m} \frac{K_{ijk} b_{ijk}}{B_{ij}} \qquad (K_z)_{ij} = \frac{B_{ij}}{\displaystyle\sum_{k=1}^{m} \frac{b_{ijk}}{K_{ijk}}} \qquad B_{ij} = \sum_{k=1}^{m} b_{ijk}$$

where *i* represents column, *j* represents row and *k* represents layer number. Vertical anisotropy for each hydrogeologic unit or model layer may be computed using above equations, if sufficient stratigraphic information is available. The anisotropy ratio may also be estimated during model calibration. The thickness and vertical hydraulic conductivity of stream and lake sediments are required for estimating seepage. These values may be obtained from field measurements or during model calibration.

Monitoring of Groundwater Levels

To obtain data on the depth and configuration of the water table, the direction of groundwater movement, and the location of recharge and discharge areas, a network of observation wells and/or piezometers has to be established. The objectives of the groundwater level monitoring are to:

- detect impact of groundwater recharge and abstractions,
- monitor the groundwater level changes,
- assess depth to water level,
- detect long term trends,
- compute the groundwater resource availability,
- assess the stage of development,
- design management strategies at regional level.

The water table reacts to various recharge and discharge components that characterize a groundwater system and is therefore constantly changing. Important in any drainage investigation are the (mean) highest and the (mean) lowest water table positions, as well as the mean water table of a hydrological year. For this reason, water level measurements should be made at frequent intervals for at least a year. The interval between readings should not exceed one month, but a fortnight may be better. All measurements should, as far as possible, be made on the same day because this gives a complete picture of the water table.

Monitoring of Groundwater Quality

For various reasons, a knowledge of the groundwater quality is.required. These are:

- Any lowering of the water table may provoke the intrusion of salty groundwater from adjacent areas, or from the deep underground, or from the sea. The drained area and its surface water system will then be charged daily with considerable amounts of dissolved salts;

- The disposal of the salty drainage water into fresh-water streams may create environmental and other problems, especially if the water is used for irrigation and/or drinking;

- In arid and semi-arid regions, soil salinization is directly related to the depth of the groundwater and to its salinity;

- Groundwater quality dictates the type of cement to be used for hydraulic structures, especially when the groundwater is rich in sulphates.

Groundwater is sampled to assess its quality for a variety of purposes. Whatever the purpose, it can only be achieved if results are representative of actual site conditions and are interpreted in the context of those conditions. Substantial costs are incurred to obtain and analyze samples. Field costs for drilling, installing, and sampling monitoring wells and laboratory costs for analyzing samples are not trivial. The utility of such expenditures can be jeopardized by the manner in which reported results are interpreted as well as by problems in how samples were obtained and analyzed. Considerable attention has been given to standardizing procedures for sampling and analyzing groundwater. Although following such standard procedures is important and provides a necessary foundation for understanding results, it neither guarantees that reported results will be representative nor necessarily have any real relationship to actual site conditions.

Comprehensive data analysis and evaluation by a knowledgeable professional should be the final quality assurance step, it may indeed help to find errors in field or laboratory work that went otherwise unnoticed, and provides the best chance for real understanding of the meaning of reported results. To facilitate interpretation, the following steps should be included:

1. Collection, analysis and evaluation of background data on regional and site-specific geology, hydrology and potential anthropogenic factors that could influence groundwater quality and collection of background information on the environmental chemistry of the analytes of concern.

2. Planning and carrying out of field activities using accepted standard procedures capable of producing data of known quality.

3. Selection of a laboratory to analyze groundwater samples based on careful evaluation of laboratory qualifications.

4. The use of appropriate quality control/quality assurance (QC/QA) checks of field and laboratory work (including field blank, duplicate, and performance evaluation samples).

5. Comprehensive interpretation of reported analytical data by a knowledgeable professional. The analytical data must be accompanied by appropriate QC/QA data, be cross-checked using standard water quality checks and relationships where possible, and be correlated with information on regional and site-specific geology and hydrology, environmental chemistry, and potential anthropogenic influences.

The objectives of the water quality monitoring network are to:

- establish the bench mark for different water quality parameters, and compare the different parameters against the national standards,
- detect water quality changes with time,
- identify potential areas that show rising trend,
- detect potential pollution sources,
- study the impact of land use and industrialization on groundwater quality,
- Data collection for the reporting year.

The frequency of sampling required in a groundwater-quality monitoring program is dictated by the expected rate of change in the concentrations of chemical constituents and the physio-chemical properties of the water being measured. Groundwater moves slowly, perhaps only a few centimeters to a few decimeters per day, so that day-to-day fluctuations in concentrations of constituents and properties at a point (or well) commonly are too small to be detected. For monitoring concentrations of major ions and nutrients, and values of physical properties of groundwater, twice yearly sampling should be sufficient, and by varying the season selected for sampling, conditions during all the seasons could be documented over a 2-year cycle. A second group of constituents, trace inorganic and organic compounds, could be adequately monitored by collecting samples once every 2 years from wells in background areas (those areas unaffected by human activities), but more frequent sampling should be considered if the types and conditions of any upgradient sources of these compounds are changing.

Consideration of several factors suggests that monitoring of groundwater quality should be a long-term activity. Not only does the structure of the program described herein mandate long-term monitoring, but also the scales over which groundwater quality is likely to fluctuate are long. Because of the slow rate of groundwater movement, any changes in factors that affect the quality of the water in recharge areas can take a long time to be reflected in surface water bodies that are discharge areas for the groundwater. The duration of a groundwater-quality monitoring program also is affected by the time scale of changes in the source area for the chemical constituents of interest.

4.0   PROCESSING OF GROUNDWATER DATA

Before any conclusions can be drawn about the cause, extent and severity of an area's groundwater related problems, the raw groundwater data on water levels and water quality have to be processed. They then have to be related to the geology and hydrogeology of the area. The results, presented in graphs, maps, and cross-sections, will enable a diagnosis of the problems. We shall assume that such basic maps as topographic, geological, and pedological maps are available.

The following graphs and maps have to be prepared that are discussed hereunder:

- ❖ Groundwater hydrographs;
- ❖ Water table-contour map;
- ❖ Depth-to-water table map;
- ❖ Water table-fluctuation map;
- ❖ Head-differences map;
- ❖ Groundwater-quality map.

It must be emphasized that a proper interpretation of groundwater data, hydrographs, and maps requires a coordinate study of a region's geology, soils, topography, climate, hydrology, land use, and vegetation. If the groundwater conditions in irrigated areas are to be properly understood and interpreted, cropping patterns, water distribution and supply, and irrigation efficiencies should be known too.

Hydrological Information System (HIS) developed under Hydrology Project provides easy access to different variety of data. The monitoring data are systematically organized in the HIS data base, including:

- well inventory
- exploratory drilling
- pumping test data
- logging
- water level
- water quality
- rainfall data
- meteorological data

Groundwater Hydrographs

When the amount of groundwater in storage increases, the water table rises; when it decreases, the water table falls. This response of the water table to changes in storage can be plotted in a hydrograph (Figure 1). Groundwater hydrographs show the water-level readings, converted to water levels below ground surface, against their corresponding time. A hydrograph should be plotted for each observation well or piezometer. It is important to know the rate of rise of the water table, and even more important, that of its fall. If the groundwater is not being recharged, the fall of the water table will depend on:

- transmissivity of the water-transmitting layer, KH;
- storativity of this layer, S;
- hydraulic gradient, dh/dx.

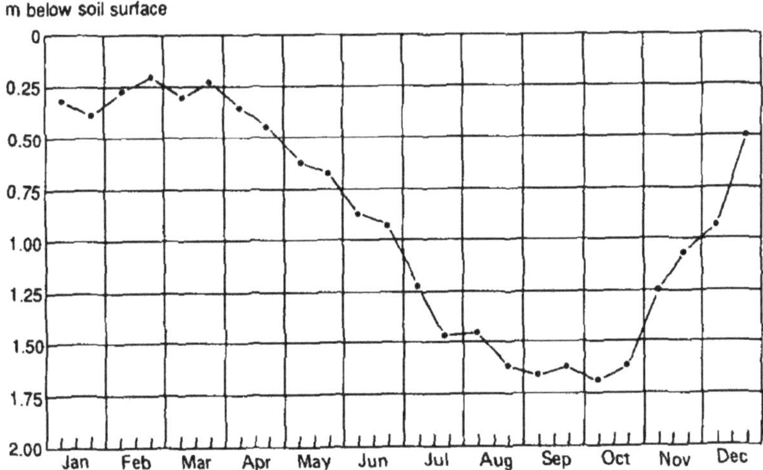

Figure 1:   **Hydrograph of a Water Table Observation Well**

After a period of rain (or irrigation) and an initial rise in groundwater levels, they then decline, rapidly at first, and then more slowly as time passes because both the hydraulic gradient and the transmissivity decrease. The graphical representation of the water table decline is known as the natural recession curve. It can be shown that the logarithm of the water table height decreases linearly with time. Hence, a plot of the water table height against time on semi-logarithmic paper gives a straight line. Groundwater recession curves are useful in studying changes in groundwater storage and in predicting future groundwater levels.

The groundwater hydrographs of all the observation points should be systematically analyzed. A comparison of these hydrographs enables us to distinguish different groups of observation wells. Each well, belonging to a certain group, shows a similar response to the recharge and discharge pattern of the area. By a similar response, we mean that the water level in these wells starts rising at the same time, attains its maximum value at the same time, and after recession starts, reaches its minimum value at the same time. The amplitude of the water level fluctuation in the various wells need not necessarily be exactly the same, but should show a great similarity. Areas where such wells are sited can then be regarded as hydrological units (i.e. sub-areas in which the watertable reacts to recharge and discharge everywhere in the same way).

Groundwater hydrographs also offer a means of estimating the annual groundwater recharge from rainfall. This, however, requires several years of records on rainfall and water tables. An average relationship between the two can be established by plotting the annual rise in water table against the annual rainfall (Figure 2). Extending the straight line until it intersects the abscissa gives the amount of rainfall below which there is no recharge of the groundwater. Any quantity less then this amount is lost by surface runoff and evapotranspiration.

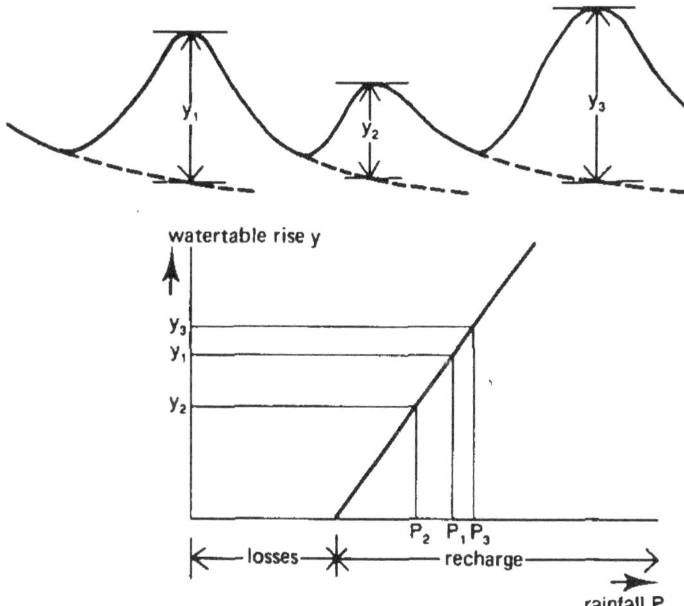

Figure 2:   Relationship between Annual Groundwater Recharge and Rainfall

Water Table-Contour Map

A water table-contour map shows the elevation and configuration of the water table on a certain date. To construct it, we first have to convert the water-level data from the form of depth below surface to the form of water table elevation (= water level height above a datum plane, e.g. mean sea level). These data are then plotted on a topographic base map and lines of equal water table elevation are drawn. A proper contour interval should be chosen, depending on the slope of the water table. For a flat water table, 0.25 to 0.50 m may suit; in steep water table areas, intervals of 1 to 5 m or even more may be needed to avoid overcrowding the map with contour lines. The topographic base map should contain contour lines of the land surface and should show all natural drainage channels and open water bodies. For the given date, the water levels of these surface waters should also be plotted on the map. Only with these data and data on the land surface elevation can water table contour lines be drawn correctly (Figure 3).

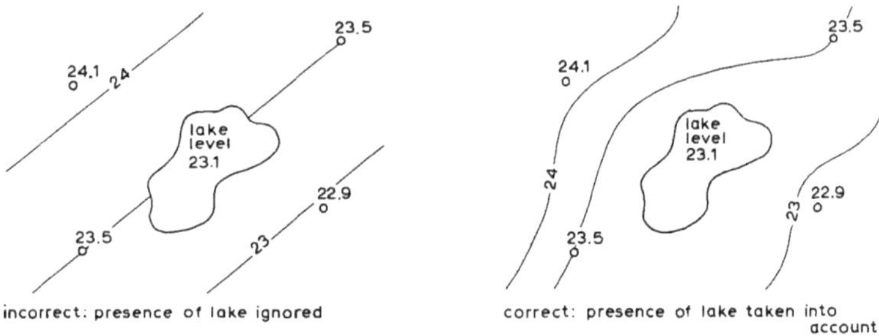

Figure 3:   **Example of Water Table-Contour Lines**

To draw the water table-contour lines, we have to interpolate the water levels between the observation points, using the linear interpolation method. Instead of preparing the map for a certain date, we could also select a period (e.g. a season or a whole year) and calculate the mean water table elevation of each well for that period. This has the advantage of smoothing out local or occasional anomalies in water levels. A water table-contour map is an important tool in groundwater investigations because, from it, one can derive the gradient of the water table (dh/dx) and the direction of groundwater flow, which is perpendicular to the water table-contour lines.

For a proper interpretation of a water table-contour map, one has to consider not only the topography, natural drainage pattern, and local recharge and discharge patterns, but also the subsurface geology. More specifically, one should know the spatial distribution of permeable and less permeable layers below the water table. For instance, a clay lens impedes the downward flow of excess irrigation water or, if the area is not irrigated, the downward flow of excess rainfall. A groundwater mound will form above such a horizontal barrier (Figure 4).

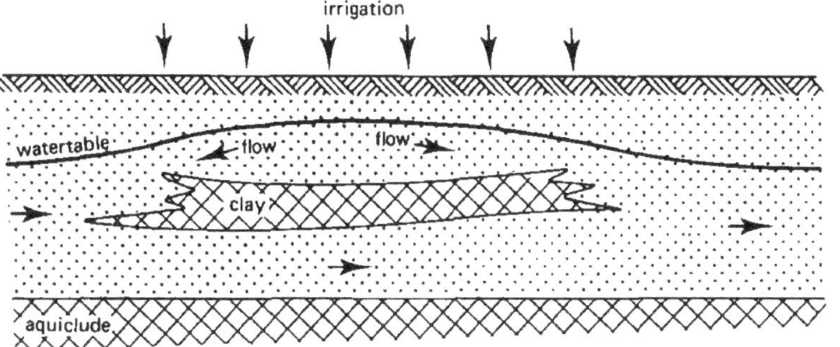

Figure 4: **A Clay Lens under an Irrigated Area impedes the Downward Flow of Excess Irrigation Water**

Water table-contour maps are graphic representations of the hydraulic gradient of the water table. The velocity of groundwater flow (v) varies directly with the hydraulic gradient (dh/dx) and, at constant flow velocity, the gradient is inversely related to the hydraulic conductivity (K), or v = -K(dh/dx) (Darcy's law). This is a fundamental law governing the interpretation of hydraulic gradients of water tables. Suppose the flow velocity in two cross-sections of equal depth and width is the same, but one cross-section shows a greater hydraulic gradient than the other, then its hydraulic conductivity must be lower. A steepening of the hydraulic gradient may thus be found at the boundary of fine-textured and coarse-textured material, or at a fault where the thickness of the water bearing layers changes abruptly.

Depth-to-Water Table Map

A depth-to-water table map or isobath map, as these names imply, shows the spatial distribution of the depth of the water table below the land surface. It can be prepared in two ways. The water level data from all the observation wells for a certain date should first be converted to water levels below land surface (the reference point from which the readings are taken needs not necessarily to be the land surface). One then plots the transformed data on the topographical base map near each observation point and draws isobaths or lines of equal depth to groundwater. A suitable contour interval may be 50 cm (Boonstra, 1988). Another way of preparing an isobath map is made by superimposing a water table contour map for a special date on the topographical map showing contour lines of the land surface. From the two families of contour lines, read the differences in elevation at contour intersections, plot these data on a clean topographical map and draw the isobaths.

According to the observation results for a year, the map drawn using the highest water table levels indicates the highest water table levels for a year in irrigated lands. The regions of map where the groundwater level is between 0-2 m depicts the area having drainage problems. On the other hand, based on measurement results for a year, the map drawn using the lowest water table levels indicates to which extent the groundwater falls in a year in irrigated area. The section where the water table level is between 0-1 m determines the areas in which groundwater exists in the root-zone throughout a year. In these areas, farm drainage systems also need to be established. The map drawn based on water table measurements done in the month of most intensive irrigation indicates to which extent irrigation activities influence the water table. Due to most intensive month in terms of irrigation application differing from one irrigation scheme to another, this month is defined as month in which most of water is released to the system.

A variety of factors must be considered if one is to interpret a depth-to-groundwater or isobath map properly. Shallow water tables may occur temporarily, which means that the natural groundwater runoff cannot cope with an incidental precipitation surplus or irrigation percolation. They may occur (almost) permanently because the inflow of groundwater exceeds the outflow, or groundwater outflow is lacking as in topographic depressions. The depth and shape of the first impermeable layer below the water table strongly affect the height of the water table. To explain differences and variations in the depth to water table, one has to consider topography, surface and subsurface geology, climate, direction and rate of groundwater flow, land use, vegetation, irrigation, and the abstraction of groundwater by wells.

Water Table-Fluctuation Map

A water table-fluctuation map is a map that shows the magnitude and spatial distribution of the change in water table over a period (e.g. a season or a whole hydrological year). Using such graphs, we calculate the difference between the highest and the lowest water table height (or preferably the difference between the mean highest and the mean lowest water table height for the two seasons). We then plot these data on a topographic base map and draw lines of equal change in water table, using a convenient contour interval. A water table-fluctuation map is a useful tool in the interpretation of drainage problems in areas with large water table fluctuations, or in areas with poor natural drainage (or upward seepage) and permanently high water tables (i.e. areas with minor water table fluctuations).

The water table in topographic highs is usually deep, whereas in topographic lows it is shallow. This means that on topographic highs there is sufficient space for the water table to change. This space is lacking in topographic lows where the water table is often close to the surface. Water table fluctuations are therefore closely related to depth to groundwater. Another factor to consider in interpreting water table-fluctuation maps is the drainable pore space of the soil. The change in water table in fine-textured soils will differ from that in coarse-textured soils, for the same recharge or discharge.

Head-Differences Map

A head-differences map is a map that shows the magnitude and spatial distribution of the differences in hydraulic head between two different soil layers. We calculate the difference in water level between the two piezometers, and plot the result on a map. After choosing a proper contour interval (e.g. 0.10 or 0.20 m), we draw lines of equal head difference. Another way of drawing such a map is to superimpose a water table-contour map on a contour map of the piezometric surface of the underlying layer. We then read the head differences at contour line intersections, plot these on a base map, and draw lines of equal head difference. The map is a useful tool in estimating upward or downward seepage.

The difference in hydraulic head between the shallow and the deep groundwater is directly related to the hydraulic resistance of the low-permeable layer(s). Because such layers are seldom homogeneous and equally thick throughout an area, the hydraulic resistance of these layers varies from one place to another. Consequently, the head difference between shallow and deep groundwater varies. Local 'leaks' in low-permeable layers may result in anomalous differences in hydraulic heads. The hydraulic resistance is especially of interest when one is defining upward seepage or natural drainage or the possibilities for tubewell drainage.

Groundwater-Quality Map

A groundwater-quality map e.g. an electrical-conductivity (EC) map is a map that shows the magnitude and spatial variation in the salinity of the groundwater. The EC values of all representative wells (or piezometers) are used for this purpose. Groundwater salinity varies not only horizontally but also vertically; a zonation of groundwater salinity is common in many areas (e.g. in delta and coastal plains, and in arid plains). It is therefore advisable to prepare an electrical-conductivity map not only for the shallow groundwater but also for the deep groundwater.

In electrical conductivity maps, critical groundwater salinity is taken as 5000 micro-mhos/cm, although it changes according to species of the crop to be grown. However, it must be examined whether salt accumulation risks occur in the root-zone through plotting of all the areas having that salinity level on the critical highest depth-to-water table map. By plotting all the EC values on a map, lines of equal electrical conductivity (equal salinity) can be drawn. Preferably the following limits should be taken: less than 100 micro-mhos/cm, 100 to 250; 250-750; 750 to 2500; 2500 to 5000; and more than 5000. Other limits may, of course, be chosen depending on the salinity found in the waters. Other types of groundwater-quality maps can be prepared by plotting different quality parameters e.g. Sodium Adsorption Ratio (SAR) values.

Spatial variations in groundwater quality are closely related to topography, geological environment, direction and rate of groundwater flow, residence time of the groundwater, depth to water table, and climate. Topographic highs, especially in the humid zones, are areas of recharge if their permeability is fair to good. The quality of groundwater in such areas almost resembles that of rainwater. On its way to topographic lows (areas of discharge), groundwater becomes more mineralized because of the dissolution of minerals. Although the water may be still fresh in discharge areas, its electrical conductivity can be several times higher than in recharge areas.

The groundwater in the lower portions of coastal and delta plains may be brackish to extremely salty, because of sea-water encroachment and the marine environment in which all or most of the mass of sediments was deposited. Their upper parts, which are usually topographic highs, are now-a-days recharge areas and consequently contain fresh groundwater.

In the arid and semi-arid zones, shallow water table areas, as can be found in the lower parts of alluvial fans, coastal plains, and delta plains, may contain very salty groundwater because of high rates of evaporation. Irrigation in such areas may contribute to the salinity of the shallow groundwater through the dissolution of salts accumulated in the soil layers. Sometimes, however, irrigated land can have groundwater of much better quality than adjacent non-irrigated land. Because of the irrigation percolation losses, the water table under the irrigated land is usually higher than in the adjacent non-irrigated land. Consequently, there is a continuous transport of salt-bearing groundwater from the irrigated to the non-irrigated land. This causes the water table in the non-irrigated land to rise to close to the surface, where evapotranspiration further contributes to the salinization of groundwater and soil.

Interpretation of Hydraulic Head and Groundwater Conditions

Groundwater flow direction: Measurements of hydraulic head, normally achieved by the installation of a piezometer or well point, are useful for determining the directions of groundwater flow in an aquifer system. In Figure 5(a), three piezometers installed to the same depth enable the determination of the direction of groundwater flow and, with the application of Darcy's law, the calculation of the horizontal component of flow. In Figure 5(b), two examples of piezometer nests are shown that allow the measurement of hydraulic head and the direction of groundwater flow in the vertical direction to be determined either at different levels in the same aquifer formation or in different formations.

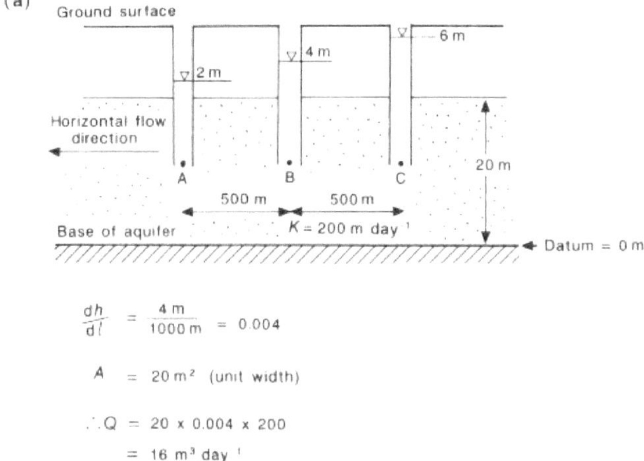

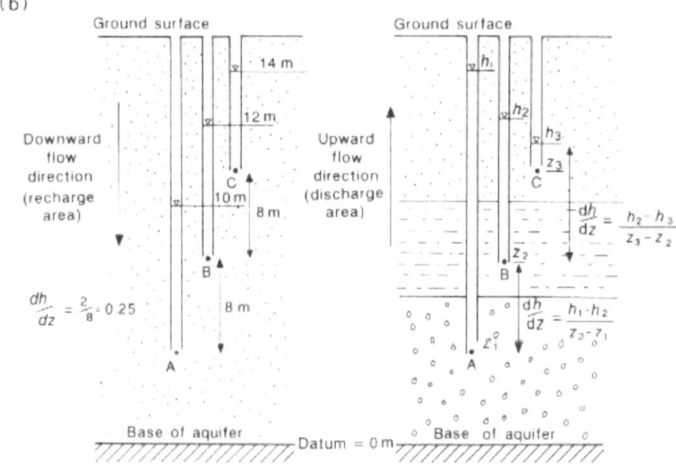

Figure 5: Determination of groundwater flow direction and hydraulic head gradient from piezometer measurements for (a) horizontal flow and (b) vertical flow. The elevation of the water level indicating the hydraulic head at each of the points A, B and C is noted adjacent to each piezometer.

5.0   INTERPOLATION OF FIELD DATA BY KRIGING

Assigning field data as input to a groundwater model is difficult because the model requires values for each grid node or cell and field data are typically sparse. To define spatial variation of a parameter over the study area, interpolation of measured data points may need to be carried out using suitable interpolation techniques like inverse distance to a power method, least square fitting of a polynomial, kriging etc.

Kriging: The most often used method for interpolation is kriging. It is a statistical interpolation method that chooses the best linear unbiased estimate for the variable in question (DeMarsily, 1986). The variable is assumed to be a random function with some kind of correlation (structure) in its spatial distribution, that is defined by a variogram (Figure 6). A variogram is a measure of the change in the variable with changes in distance.

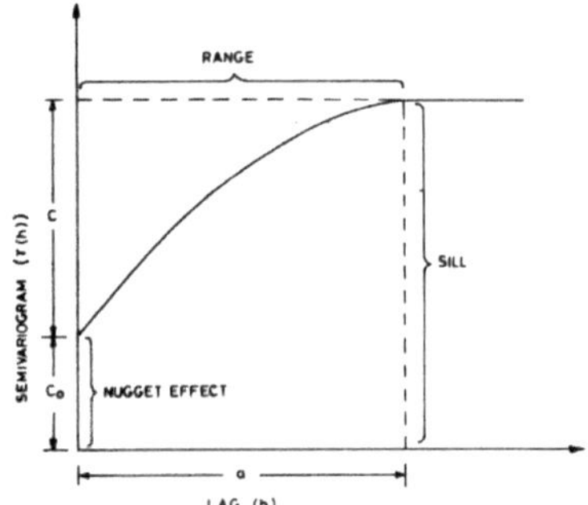

Figure 6: Plot of Variogram

Higher correlation is expected between points measured at small separation distances. For example, hydraulic head, transmissivity, permeability, thickness of a layer, storage coefficient, rainfall, effective recharge, etc., are all functions of space and are very often highly variable. This spatial variability is, in general, not purely random; if measurements are made at two different locations, the closer the measurements points are to each other, the closer are the measured values.

For the purpose of analysis, the spatial continuity is expressed as a variogram, which is a plot of the semi-variance (γ) against the separation distance (h). The semi-variance is defined as half the mean of squares of the differences of the recorded values at a pair of stations spaced at a given distance. According to the definition of γ, its value at h equal to zero is half the variance of a data from itself, which is zero. Thus, the variogram has to pass through the origin. Further, for a distance less than the minimum distance between the observation points, γ is not defined. The minimum distance may be quite small in comparison with the total range of h, and γ corresponding to this distance may not be close to zero. Since, the variogram has to pass through the origin, this leads to a nugget effect, which is basically a vertical rise of the plotted curve at the origin (Figure 6). The value of γ rises gradually upto a limiting distance beyond which it becomes a constant. This distance is known as the range of the variable. The maximum value of γ reached at the range is known as the sill. The range can be viewed as the effective neighbourhood within which the continuity of the variable holds.

The structural information as represented in the variogram is then fitted with a model. The commonly used models are linear, spherical, exponential and Gaussian. These are illustrated in Figure 7.

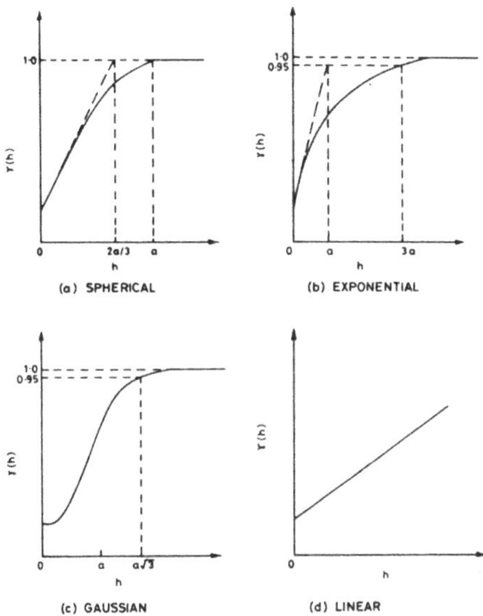

Figure 7:   Theoretical Models of Variogram

The model fitting involves choosing an appropriate model and estimating its parameters, such as the range and nugget. Thus, kriging is the process of estimating the value of a spatially distributed variable from adjacent values while considering the interdependence expressed in the fitted variogram model.

Kriging differs from other interpolation methods because it considers the spatial structure of a variable, and also preserves the field value at measurement points (unlike some other interpolation schemes such as least square fitting of a polynomial). Irrespective of the method used to assign parameter values to nodal points of grid, care should be taken to check that the resulting parameter distribution is reasonable in hydrogeologic terms and that the values fall within appropriate range for the geologic setting.

6.0   GIS FOR GROUNDWATER STUDIES

For handling groundwater data, the GIS-technology is aptly suited, for the following main reasons:

1. *Concurrent handling of locational and attribute data:* In groundwater studies, one has to deal with information comprising locational data (where it is?) and attribute data (what it is?). GIS packages have the unique capability to handle locational and attribute data; such a capability is not available in other groups of packages (Figure 8).

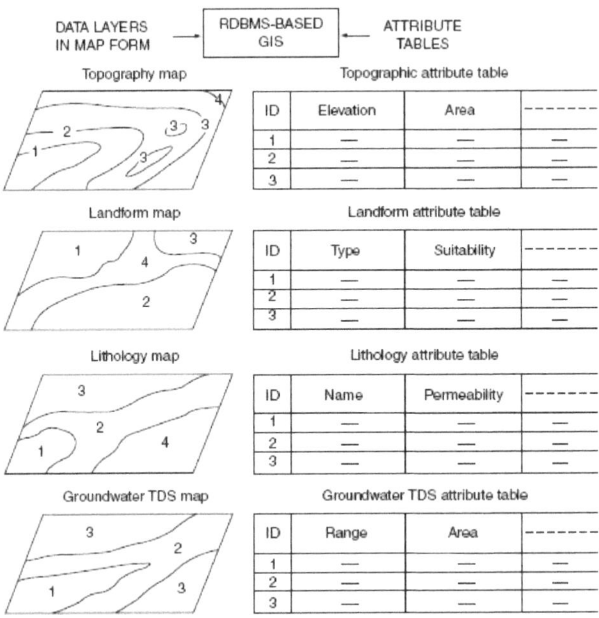

Figure 8:   Schematic Representation of GIS-Working

2. *Variety of data:* Groundwater investigations often comprise diverse forms and types of data, such as: (a) topographic contour maps, (b) landform maps, (c) lithological maps, (d) structural geological maps, (e) isobath map (contour map of equal depth of water-table), (f) isogram (isocone) maps depicting groundwater characteristics by contours of equal concentration of dissolved solids (TDS) or ions, (g) drainage density and other geomorphic maps, (h) tables of various observations and data sets, and (i) point data, say locations and water-levels in observations wells, or spring discharge etc. In these, some of the variables

are of continuous type, e.g. TDS content, water-level data etc., and some others are of categorical type, such as low/medium/high drainage density, or gravel/marble/granite lithology. It is essential to integrate the spatial information for coherent and meaningful interpretation, and to avoid compartmentalization of data. GIS offers technological avenues for integrating the variety of data sets in both qualitative and quantitative terms, hitherto not available through any other route.

3. *Flexibility of operations and concurrent display:* Modern GIS packages are endowed with numerous functions for computing, searching for and classifying data, which allow processing and analysis of spatial information in a highly flexible manner and concurrent display, interactively.

4. *Speed, time and costs of processing:* Advances in micro-electronics and computer technology have made it possible that modern GIS can store, process and analyze large volumes of data which otherwise would be too expensive, tedious and time-consuming to carry out by other methods.

5. *Higher accuracy and repeatability of results:* The technique being digital computer-based, yields higher accuracy, in comparison to manual cartographic products. The results are amenable to re-checking and confirmation. In a typical hydrogeological investigation, the sources of data could be satellite or aerial sensing, field surveys, geochemical laboratory analyses, geophysical exploration data, etc. and may be available in the form of maps, profiles, point data, tables, lists etc. If GIS methodology is not used, then integrating such a variety of data sets would involve elaborate manual exercises in order to deduce the relevant information.

6.1 Arc Hydro - Geographic Data Model

The Arc Hydro groundwater data model is a geographic data model for representing spatial and temporal groundwater information within a geographic information system (GIS). The data model is a standardized representation of groundwater systems within a spatial database that provides a public domain template for GIS users to store, document and analyze commonly used spatial and temporal groundwater data sets. It includes two-dimensional (2D) and three-dimensional (3D) object classes for representing aquifers, wells, and borehole data and 3D geospatial context in which these data exist. The framework data model also includes tabular objects for representing temporal information such as water levels and water quality samples that are related with spatial features.

6.2 Geostatistical Analysis using ArcGIS

GIS is designed to support a range of different kinds of analysis of geographic information: techniques to examine and explore data from a geographic perspective, to develop and test models, and to present data in ways that lead to greater insight and understanding. A linkage between GIS and spatial data analysis is considered to be an important aspect in the development of GIS into a research tool to explore and analyze spatial relationships. The GIS methodology for the spatial analysis of the groundwater levels (as illustrated in Figure 9) involves the following steps:

(a) Exploratory spatial data analysis (ESDA) using ArcGIS software for the water level to study the following:

> Data distribution.
> Global and local outliers.
> Trend analysis.

(b) Spatial interpolation for water level data using ArcGIS software, while ordinary kriging is applied by involving the following procedures:

> Semi-variogram and covariance modelling.
> Model validation using cross-validation.
> Surfaces generation of the groundwater level data.

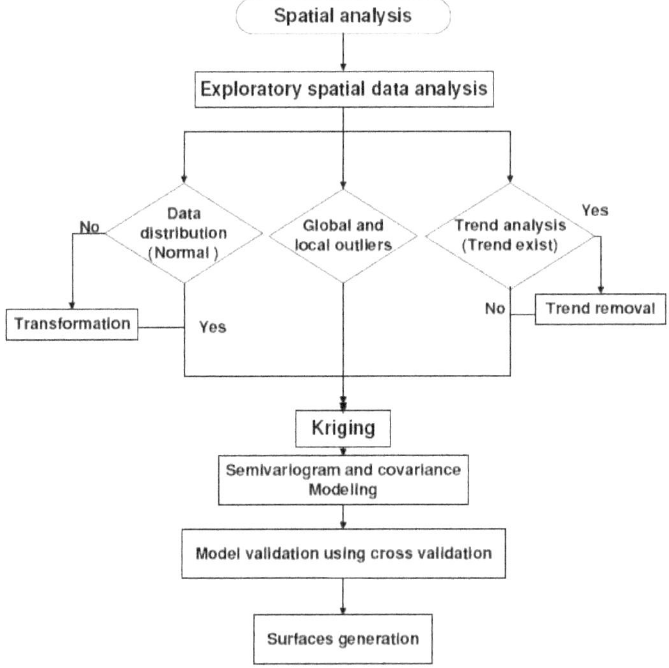

Figure 9:   **Flow Chart of the Geostatistical Analysis Steps**

There are some methods of spatial modelling that work better if the experimental distribution of available data is close to normal one, therefore it is necessary to check for normality before performing any spatial modelling. Transformations are necessary to drive the data to normal distribution in case of non-normality, where several transformations including Box–Cox also known as power transformations, arcsine, and logarithmic one, can be used to make the data more normally distributed. The groundwater levels data are plotted in the histogram, where groundwater level data is not normally distributed.

Trend is a surface that may be made up of two main components: a fixed global trend and random short-range variation. The global trend is sometimes referred to as the fixed mean structure. Somewhat different approach to ESDA for continuous data represented as a point set with z-values is to examine whether any simple trends are present. The spatial trend can be further analyzed using a three-dimensional perspective using ArcGIS geostatistical analyst trend analysis tool.

A global outlier is a measured sample point that has a very high or a very low value relative to all of the values in a dataset. For example, if 99 out of 100 points have value between 300 and 400, but the 100th point has a value of 750, the 100th point may be a global outlier. Local outlier is a measured sample point that has a value that is within the normal range for the entire dataset, but when compared to the surrounding points, it is unusually high or low.

Kriging is divided into two distinct tasks: quantifying the spatial structure of the data and producing a prediction. Quantifying the structure, known as variography, is where spatial-dependence model fit the data. For making a prediction for an unknown value for a specific location, kriging will use the fitted model from variography, spatial data configuration, and values of the measured sample points around the prediction location. Variography is the process of estimating the theoretical semi-variogram.The semi-variogram and covariance function quantify the assumption that things nearby tend to be more similar than things that are farther apart. They both measure the strength of statistical correlation as a function of distance.

Validation should be carried out before producing the final surface, where it helps in making an informed decision as to which model provides the best predictions. The most popular methods for verifying predictions are cross-validation and validation provided in ArcGIS Geostatistical Analyst. If the mean prediction error is near zero, predictions are centred on the measurement values. The closer the predictions are to their true values, the smaller the root-mean-square prediction errors. The average root-mean-square prediction errors are computed as the square root of the average of the squared difference between observed and predicted values. For a model that provides accurate predictions, the root-mean-squared prediction error should be as small as possible. The average standard error and the mean standardized prediction error should be as small as possible. After model validation, the surface can be generated to produce the water level map.

7.0 DATABASE FOR COASTAL AQUIFER PLANNING

Given the multi-disciplinary nature of coastal aquifer studies, one of the most important elements in the overall planning approach is adequate database development and application. Data must be organized in such a way that it can be analyzed spatially, in three dimensions, as well as temporally. The long-term nature of interface movement requires that data from as far back as possible be collected. The only way to make the data available for analysis and modeling is to develop an integrated database/geographic information system (GIS). This critical, and often neglected, step of integrated database design allows users and modelers to analyze and query data, and places the data in a consistent format for model pre- and post-processing. Data elements and map coverages in the database/GIS typically needed for coastal aquifer management include:

- Well information (depth, location, aquifer designation - even if preliminary)
- Historic and projected pumping information (linked to the well information)
- Chloride sampling data (dated, linked to well locations)
- Water level data (dated, linked to well locations)
- Surface map features (roads, streams, well locations, topographic features)
- Aquifer hydrogeologic parameters (transmissivity, hydraulic conductivity, formation thickness, specific yield, storativity, others). Data may exist as discrete points or spatial contours.
- Recharge estimates, mapped as contours if spatial variation is expected
- Maps of estimated present interface locations and depths

Long-term pumping records must also be collected. These data are critical to the development of a groundwater model. Unlike the calibration of a typical groundwater model in a freshwater aquifer, the response time of the freshwater/saltwater interface to changed pressure distribution (rise in sea level, increased pumping, altered recharge) in a coastal system might well be decades, or in some cases even a century or more. A critical part of the conceptual model is the estimate of the natural position of the interface prior to pumping, and a determination of whether the pre-development position was in equilibrium, or, the aquifer is still responding to a long-term change in sea level from the last glacial period.

Due to the slow response of the interface, estimates of pumping rates over many decades must be made to test the model. Once the data have been put into a database/GIS, initial analysis can be carried out prior to modeling. Common analytical steps include examining:

- Water quality trend and spatial analyses
- Pumping analyses: monthly, seasonal, annual
- Water level and aquifer head mapping
- Chloride concentration and trend mapping

Most commercially available database software are now powerful enough to handle the data needs for even a large-scale regional aquifer management study. The key is to set up the database and the groundwater model in such a way that data can be moved from the database/GIS into the model, and model results can be transferred back to the database/GIS with relative ease.

8.0   GROUNDWATER DATA MANAGEMENT AND ANALYSIS TOOLS

Following are some of the software packages used for groundwater data management and analysis.

AQTESOLV: Design and analysis of aquifer tests including pumping tests, step-drawdown tests, variable-rate tests, recovery tests, single-well tests, slug tests and constant-head tests.

AquaChem: AquaChem is an integrated software package developed specifically for graphical and numerical analysis of geochemical data sets. It features a database of common geochemical parameters that can be customized and configured to include an unlimited number of attributes per sample. The built-in analysis tools include many common calculations used for analyzing, interpreting and comparing aqueous geochemical data. The built-in graphics include many common geochemical plotting methods such as Piper, Stiff, Durov, Radial, Schoeller, Langlier-Ludwig and more. AquaChem also includes a direct link to the popular PHREEQC program for geochemical modeling.

AquiferTest Pro: Graphical analysis and reporting of pumping test and slug test data.

AquiferWin32: A software system for analysis and display of aquifer test results.

Arc Hydro Groundwater: Arc Hydro Groundwater (AHGW) tools include Groundwater Analyst, MODFLOW Analyst, and Subsurface Analyst. It can access and visualize your groundwater, time series, and geologic data within ArcGIS.

ChemStat: RCRA compliant statistical analysis of groundwater data. It includes most methods described in 1989 and 1992 USEPA statistical guidance documents.

EnviroInsite: EnviroInsite is a desktop, groundwater visualization package for analysis and communication of spatial and temporal trends in multi-analyte, environmental groundwater data. It facilitates the generation of data queries to generate time history graphs, pie charts, radial diagrams, data tables and dot-plots in plan, on vertical profiles, and in 3D views.

Environmental Visualization System (EVS): Unites advanced gridding, geostatistical analysis, and fully three-dimensional visualization tools into a software system developed to address the needs of all earth science disciplines. The graphical user interface is integrated with modular analysis and graphics routines. The more advanced versions allow these modules to be customized and combined. The software can be used to analyze all types of analytes and geophysical data in any environment (e.g. soil, groundwater, surface water, air, etc.). It includes integrated geostatistics.

EQWin Data Manager: EQWin Data Manager is a database application used to validate, store, analyze, and report on environmental data. Using either MS Access or SQL Server databases, EQWin manages many frequently monitored entities, including ground/surface water, soil, air, and meteorological data. EQWin is integrated with Microsoft Excel as its default import/reporting tool.

GW Contour: Data interpolation and contouring program for groundwater professionals that also incorporates mapping velocity vectors and particle tracks.

GW-Base: GW-Base was developed for expedient evaluation of groundwater information for use in water management, monitoring, investigation and remediation of damage cases. It combines a powerful database with easy-to-use evaluation tools like contour maps, bar-charts, pie-charts, statistical tools, query options and GIS-functionalities.

HydroGeo Analyst: Groundwater and borehole data management and visualization technology.

LogPlot: Log plotting software for the environmental, petroleum, mining, and academic geoscientist.

PUMPTEST (IGWMC): The PUMPTEST program package is a menu-driven set of independently run programs. It includes three different methods to analyze pumping test data: the time-drawdown method (JACOBFIT); distance-drawdown method (DISTANCE); and recovery method (RECOVERY).

RockWorks: Geological data management, analysis and visualization.

Strater: Strater is a well log and borehole plotting software program that imports data from a multitude of sources (database files, data files, LAS files, ODBC, and OLE DB data sources). Strater provides innumerable ways to graphically display the data. All the logs are fully customizable.

WinLoG: WinLoG can be used to quickly create, edit and print geotechnical, environmental, mining, and oil & gas borehole and well logs. The graphical windows interface displays the log as it is changed and shows exactly how the log will look when it is printed.

9.0   REVIEW OF GROUNDWATER SCENARIO

Review of Groundwater Level Changes

> - Describe the typical long-term water level hydrographs. Show typical examples of villages showing rising water level trends and declining water level trends.
> - Describe the typical long-term water level hydrographs for typical areas (coastal areas, irrigation commands, over-exploited areas, delta areas, areas close to riverbeds).
> - Study multiple hydrographs from a number of wells within a watershed to understand the groundwater dynamics. Assess the water level changes in multi-aquifer system.
> - Delineate areas showing typical long-term water level trends (with list of villages).
> - Describe typical high frequency water level monitoring hydrographs and their significance.
> - Explain the recharge – rainfall response for different rainfall intensities.

Review of Groundwater Flow System Characteristics

- Generate water level fluctuation contour maps using data for the reporting period and the last year (pre/post monsoon) from all the monitoring wells tapping a single aquifer in the network.
- Generate water level elevation contour maps using data for the reporting period (pre/post monsoon) from all the monitoring wells tapping a single aquifer in the network.
- From the generated map, assess the gradient of groundwater flow, determine the flow path, and delineate the recharge and discharge areas. Detect any change in flow gradient or path, as compared to the previous years.
- Generate maps for all the different aquifers and assess the gradient of groundwater flow for the different aquifers. Assess nature of contact/mixing between the aquifers.
- Assess the groundwater flow through the aquifer system using supporting data (rainfall, runoff, recharge and draft).
- Generate Groundwater worthy map.

Review of the Groundwater Quality Changes

- Describe the groundwater quality monitoring network, frequency of monitoring, list of laboratories and parameters analysed.
- Show sample hydrograph depicting the changing trends in water quality for different parameters. List the parameters that show higher levels of concentration or show increasing trend.
- Describe the chemical quality of groundwater in different aquifers; assess the parameters that show higher levels of concentration or show increasing trend.
- Generate water quality contour maps for specific parameters using analysed results for the reporting period (pre/post monsoon) from all the monitoring wells tapping a single aquifer in the network.
- From the generated maps, assess the pattern of contaminant transport, delineate areas showing high concentration, identify polluting sources, if any (natural/industrial).
- Generate water quality maps and diagrams for different aquifers. Assess nature of contact/mixing of contaminants, if any, between aquifers.
- Assess the rate of dilution or increasing concentration. Study the impact of rainfall, surface water bodies, recharge and excess draft on groundwater quality.

Estimation of Groundwater Resource Availability

- Carry out groundwater resource estimation for the reporting period based on the GEC (Groundwater Resource Estimation Committee, 1997) norms.
- Identify watersheds/administrative units subjected to overexploitation, as compared to previous years.
- Identify areas that are showing heavy increase in draft.
- Estimate stage of groundwater development.
- Generate notified area map.

Recommendation for Sustainable Development of Groundwater

- ➤ Based on different analyses, list the administrative blocks showing declining and rising water levels.
- ➤ Delineate recharge and discharge areas.
- ➤ Identify the technically appropriate programs that need to be considered for containing the declining water level trend and increasing contamination, containing depletion of resources, reducing erosion and increasing recharge.
- ➤ Recommend the appropriate designs for efficient wells, artificial recharge structures/water harvesting ponds that can be taken up in different areas.
- ➤ Recommend sustainable groundwater development programs for ecolgically fragile areas like coastal/urban/ industrial areas.
- ➤ Identify specific research projects that need to be considered for tackling serious groundwater related issues.
- ➤ Recommend to administrators/planners the appropriate groundwater policies and legislation that can ensure equity and ensure groundwater sustainability.

REFERENCES

Boonstra, J. (1988). *Groundwater Survey. Groundwater Balances*. Lecture Note. Twenty - Ninth International Course on Land Drainage (1990), ILRI, Wageningen, The Netherlands.

Cheng, Alexander H.-D. And Ouazar, Driss (2004). *Coastal Aquifer Management Monitoring, Modeling, and Case Studies*, Lewis Publishers.

De Marsily, G. (1986). *Quantitative Hydrogeology*, Academic Press.

de Ridder, N. A. (1974). Groundwater Investigations, In: *Drainage Principles and Applications*, ILRI Publication 16.

DHV Consultants. *Model Yearbook – Groundwater*, Hydrology Project (India).

"Groundwater Resource Estimation Methodology - 1997". Report of the Groundwater Resource Estimation Committee, Ministry of Water Resources, Government of India, New Delhi, June 1997.

Gundogdu, K. S., Demir, A. O., Degirmenci, H., Buyukcangaz, H., Akkaya, T. *Preparation and Interpretation of Groundwater Maps using Geographical Information System (Arc/Info)*.

Moore, J. E. (1979). *Contributions of Groundwater Modelling to Planning*, Journal of Hydrology, Vol. 43, pp. 121-128.

Salah, Hamad (2009). *Geostatistical Analysis of Groundwater Levels in the South Al Jabal Al Akhdar Area using GIS*. GIS Ostrava 2009.

Singhal, B. B. S. and Gupta, R. P. (2010). *Applied Hydrogeology of Fractured Rocks*, Springer.

Zemansky, G.M. (1998). *Interpretation of Groundwater Chemical Quality Data*, Proceedings, 14th Annual Waste Testing & Quality Assurance Symposium, July 13-15, 1998, Arlington, Virginia, pp. 192-201.